Orchids of the Philippines

GIFTWRAPS AROUND THE WORLD

Introduction
Louise Feinsot

• • •

Design Consultant
Joost Elffers

Harry N. Abrams, Inc., Publishers, New York

Editor: Eric Himmel
Designer: Darilyn Lowe
ISBN 0-8109-2975-9

Published in 1990 by Harry N. Abrams, Inc., New York

A Times Mirror Company
Printed and bound in Italy

INTRODUCTION

In the vast universe of flowering plants, none have excited the imagination and stirred the senses more than orchids. Their haunting, unforgettable beauty; sensuous, exotic and often sinister forms; vibrant colors; and remarkable scents have inspired poets and painters and made them favorites of botanists, collectors, and growers around the globe for centuries.

In the Western world, the ancient Greeks were the first to write about orchids. History records that the Greek philosopher, Theophrastus (c. 372–c. 287 B.C.), a pupil of Aristotle, first used the term "orchis" to describe them in his manuscript *Enquiry into Plants.* Some three hundred years later, the word orchid was adopted by Dioscorides, a Greek physician, who attributed medicinal and aphrodisiac properties to them in his treatise *Materia Medica.* His work dominated botanical thought for sixteen centuries, helping to perpetuate myths and superstitions that came to surround these fascinating flowers.

In the East, Confucius, some 2,500 years ago, likened the orchid to the perfect or superior man, and held that its fragrance evoked the joys of friendship. References to orchids appeared in Chinese books as early as A.D. 290–390, and delicate calligraphic paintings by Chinese artists have been found dating to the eleventh and twelfth centuries. Orchid became the official name for the family of these plants when the Swedish botanist, Carl Linnaeus—known as the father of modern taxonomy—used it in his book *Species Plantarum,* published in 1753.

Interest in orchids increased dramatically after the Royal Botanic Gardens and the Royal Horticultural Society in England began to cultivate them in the early 1800s. Owning orchids became a status symbol among the privileged classes, and collectors vied for exclusive ownership of prized species. Orchid mania became widespread, and the search for new sources assumed great importance. Professional collectors traveled in droves to the tropics to hunt for new species that might attract horticultural interest.

Today one still thinks of the orchid as rare—a treasured novelty—and yet there are an estimated 35,000 species of the flowers, and new discoveries are being made each year. In addition to native and wild varieties, there are another 30,000 cultivated hybrids. But no matter how ubiquitous they are, each one is so special and evokes such extraordinary responses of pleasure that it is hard to consider them as commonly available.

The greatest number grow wild in tropical regions, particularly in areas of Southeast Asia and South America. The Philippines is one of the richest sources of orchids, with some 900 indigenous species. Because of the sheer number and variety of orchids to be found there, Western growers and collectors began coming to the Philippines on botanical expeditions starting in the early 1800s. Indeed, many of today's favorite orchids originated in the Philippines, having been discovered and carted away years ago by European collectors.

Philippine orchids are known to have enriched the bloodlines of many celebrated hybrid varieties. Among the most important are:

Aerides: An epiphytic genus with some 60 species, seven of which are found in the Philippines. These plants can grow six feet tall, producing arching stems with yellowish-white and purple, waxy flowers.

Dendrobium: A large, diverse epiphytic genus of about 1,000 species and many hybrids, with some 86 indigenous to the Philippines. Dendrobiums vary widely in size, flower, and plant structure; some are evergreen, others deciduous: their leaves fall before the plant blooms.

Paphiopedilum: A primarily terrestrial genus of 50 species, of which 13 grow in the Philippines. These orchids are known for their slipper-sandal lip and their waxy, somewhat artificial-looking flowers. There are numerous horticultural hybrids.

Phalaenopsis: Known as the moth orchid, perhaps the most beautiful orchids belong to this epiphytic genus of some 40 species. Of these, 20 species and 18 varieties are native to the Philippines. Extensively hybridized, there are many fine varieties of these exquisite flowers. Floral colors vary but are most frequently white, pink, yellow, purple, or brown.

Renanthera: Known as "Fire Orchids" because of their bright red colors. There are about 15 epiphytic species of the genus; four grow in the Philippines.

Rhynchostylis: Known as "Fox-Tail Orchids" because of the density of the blooms. A single plant can have 200 flowers. This epiphytic genus has four species; two grow in the Philippines.

Spathoglottis: A terrestrial genus of 40 species; six species and one variety grow in the Philippines. They have prominently veined fanlike leaves and bear a succession of small to rather large handsome flowers.

Trichoglottis: A genus of some 60 epiphytic species, 20 are found in the Philippines. These vinelike plants have yellowish flowers with red-brown markings.

Vanda: There are 70 epiphytic species; seven are found in the Philippines, where they are known as Waling-Waling. Hundreds of hybrids have been derived from the genus.

The majority of Philippine orchids grow in the larger land masses of the archipelago in Luzon, Mindanao, Visayas, and Palawan. The giftwraps in this book are artistic renderings of some of the best known and most beautiful of the Philippine species.

How They Grow

Orchids grow in an extraordinary range of different places. They have been found all around the world from the tropics to the Arctic Circle, from lowland plains to the snowline in mountain regions. Their habitats vary according to the species. While some are terrestrial, growing in the soil, a majority are either epiphytic, and grow non-parasitically on another plant or tree, or lithophytic, growing on the surface of rocks or beneath another growing medium. A few orchids are saprophytes—that is, they grow on dead or decaying animal or vegetable matter in the ground.

The life span of orchids is a curious phenomenon. Some live but a few minutes, while others flourish an astonishing nine months and are the longest lasting of all flowers. Orchids are unusual in still another respect—in the way they are pollinated. They do not simply attract particular kinds of pollinators, such as bees or wasps or butterflies, but are highly selective and limit themselves to members of a single variety, belonging to a single species. This pollinating process has helped the individual species preserve their discrete identity. Charles Darwin was so fascinated by the way orchids are pollinated that he wrote a famous treatise on the subject—"On the Various Contrivances by Which British and Foreign Orchids Are Fertilised by Insects."

Orchids For Everyone

The orchid remains the flower of choice for the corsage. It is still considered the most beautiful and luxurious of blooms, and millions of orchids are purchased each year for Mother's Day, anniversaries, and other festive occasions. Orchids are cultivated in many areas of the world for commercial distribution, and a wide selection are available as cut flowers year-round. Their exotic shapes and remarkable colorations add such a special dimension to flower arrangements that they are included with increasing frequency in centerpieces and other larger floral displays.

Orchids are clearly no longer just for the rich. Today they can be purchased through local florists; and collections like the Dupont Orchids at Longwood Gardens near Philadelphia, and the Fairchild Gardens in Coral Gables, Florida, which were once private are now open for public viewing.

The American Orchid Society, with headquarters at the Botanical Museum of Harvard University, has some eighteen thousand registered members and more than two hundred and fifty local affiliated orchid societies. As myths and misinformation about orchid growing have disappeared, the number of amateur growers has increased markedly and today orchids can be found among other house plants in a great many homes.

Throughout the world, orchids remain a continuing source of fascination for their wealth of bloom, beauty, and infinite variety. While no longer endowed with mystical power, their allure seems to grow stronger with each succeeding generation.

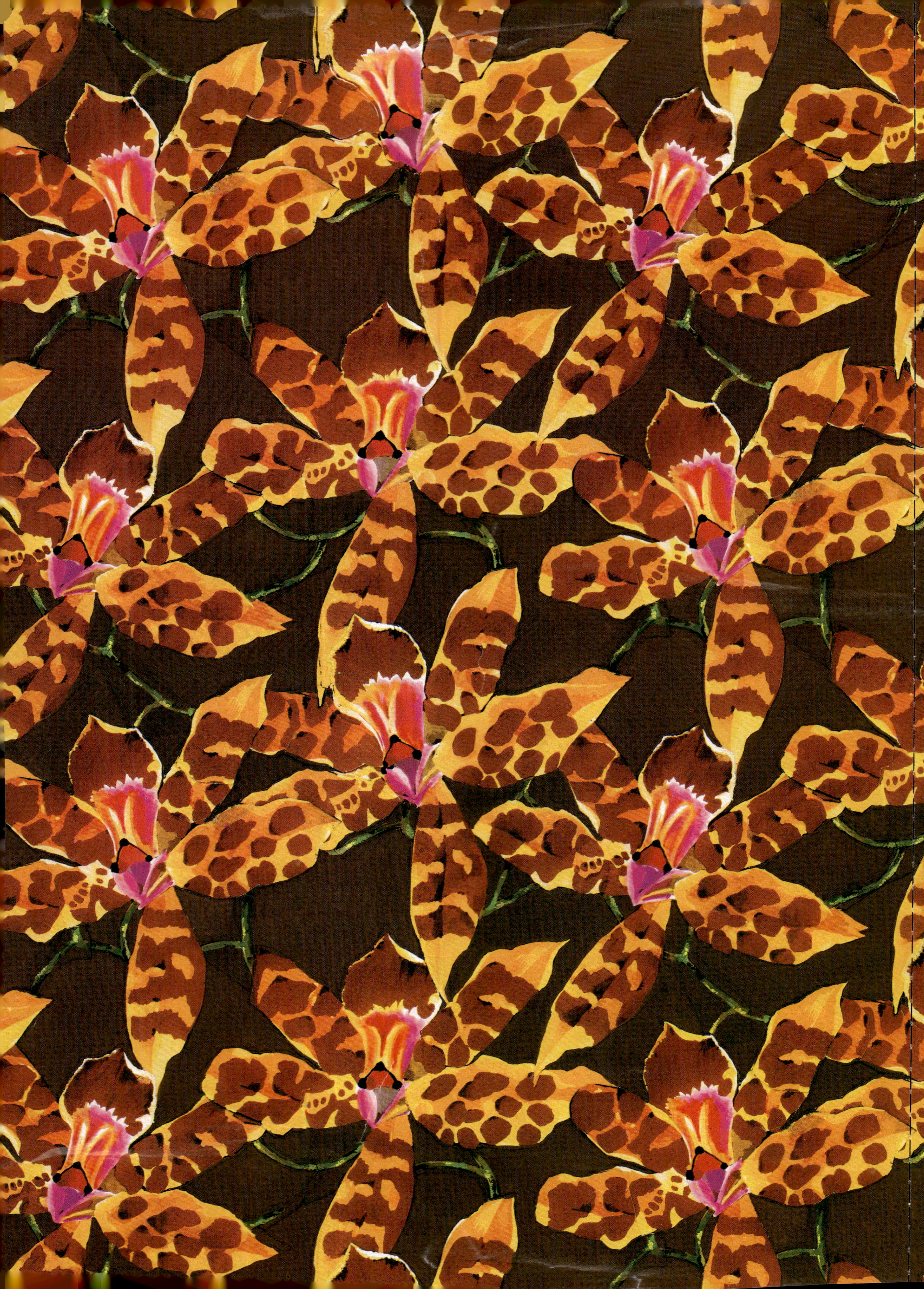